Name ____________________ Teacher ____________________

Class ____________________ Date ____________________

AF882723

Find the missing number

Solve the equations below.

$10 - 0 = \square$	$5 - 4 = \square$
$5 - 3 = \square$	$2 - 0 = \square$
$7 - 4 = \square$	$4 - 1 = \square$
$8 - 4 = \square$	$1 - 0 = \square$
$7 - 6 = \square$	$5 - 1 = \square$

NAME:

DATE:

SECTION:

SCORE:

BASIC SUBTRACTION WORKSHEET

Fill out the box with the right answer.

4 - 1 = ☐

3 - 1 = ☐

5 - 3 = ☐

4 - 2 = ☐

2 - 1 = ☐

Name

Class

Date

Score

Counting Cats and Dogs

AN ADDITION GAME

Directions: Count how many of each animal you see. Write the answers in the box provided below.

Cats

Dogs

Name: ______________________ Date: ______________

Section: ______________________ Score: ______________

FRUIT MATH

LET'S ADD THESE FRUITS UP!

Directions: How many fruits can you see? Add them all up, then write the answer in the box. Make sure to double check your answers before submitting! Perfect scores get a prize from Ms. Lucy.

NAME

CLASS

DATE

SCORE

ROCKET MATH

Solve the equations to launch the rockets!

3 x 5	8 x 7	9 x 4	6 x 3
=	=	=	=
7 x 9	1 x 8	9 x 0	5 x 4
=	=	=	=
3 x 8	8 x 6	3 x 8	9 x 2
=	=	=	=

Name: Date:

BEFORE & AFTER

123456789

Fill in the numbers that come before and after:

	1	
	5	
	9	

	12	
	16	
	19	

	22	
	27	
	29	

	31	
	37	
	39	

Name: ______________________

DIRECTIONS: TRACE THE WORDS AND NUMBERS BELOW.

three

3 3 3 3

3 3 3 three

COLOR THREE SQUARES:

CIRCLE THE THREES:

2	3	5
4	2	3
3	1	2
3	5	1
6	3	4

Name: ______________ Date: ______________

Section: ______________ Score: ______________

Kindergarten Subtraction Worksheet

Cross out the correct number of circles to solve each subtraction.

5 - 2 = ______	○○○○○
3 - 2 = ______	○○○
4 - 1 = ______	○○○○
3 - 1 = ______	○○○
5 - 1 = ______	○○○○○
4 - 2 = ______	○○○○

Count & Color

Count and color the exact number of farm animals

1

6

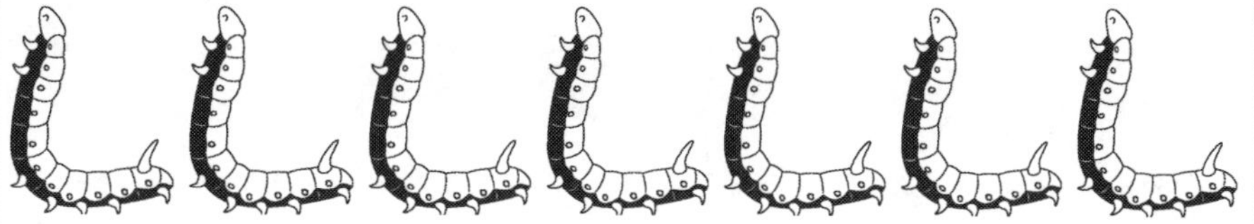

3

7

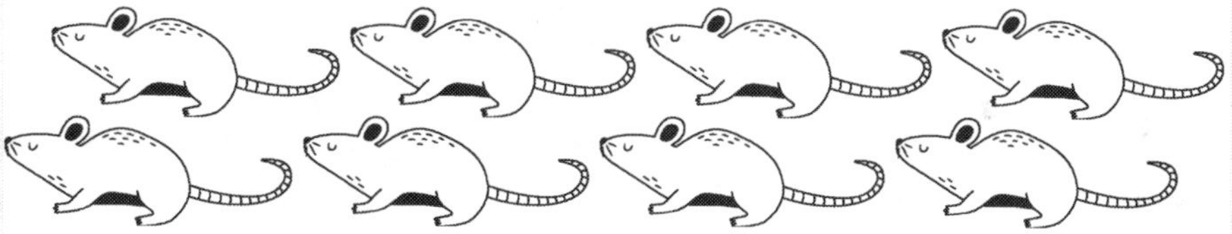

2

4

5

Count & Color

Count and color the exact number of farm animals

7	
6	
5	
10	
8	
4	
9	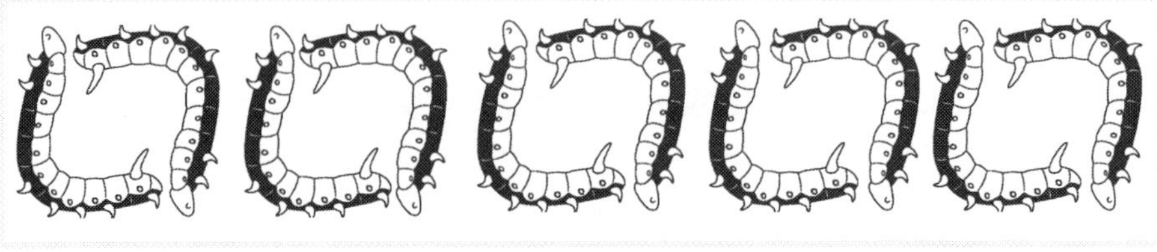

Count & Color

Count and color the exact number of farm animals

1

2

7

3

6

5

4

Name: ____________________

Cupcakes & Number Lines

Using the number line to show your work, complete the addition following the example.

1 5 10

$5 + 2 = 7$

1 5 10

$6 + 3 =$

1 5 10

$3 + 2 =$

1 5 10

$5 + 4 =$

1 5 10

$1 + 4 =$

1 5 10

$4 + 2 =$

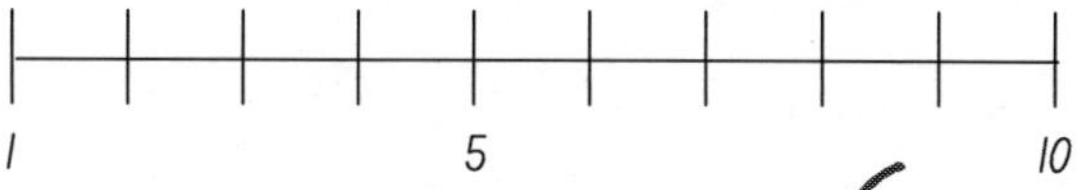

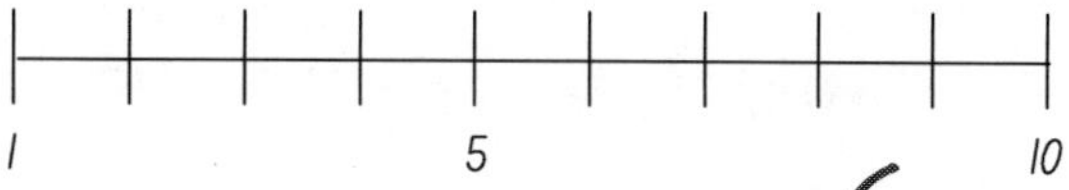

4 + 4 = 

3 + 7 = 

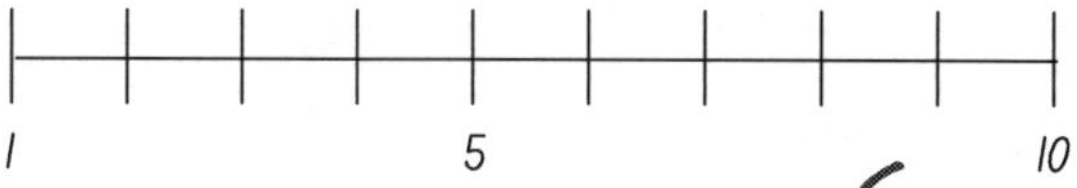

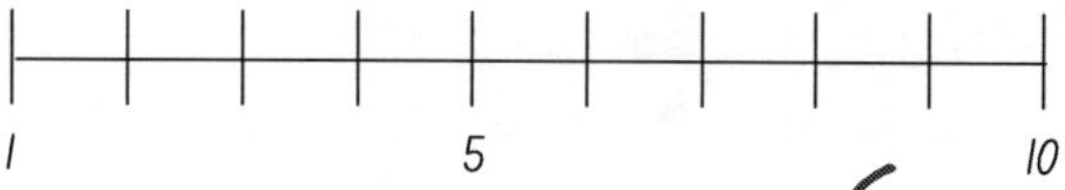

2 + 2 = 

4 + 6 = 

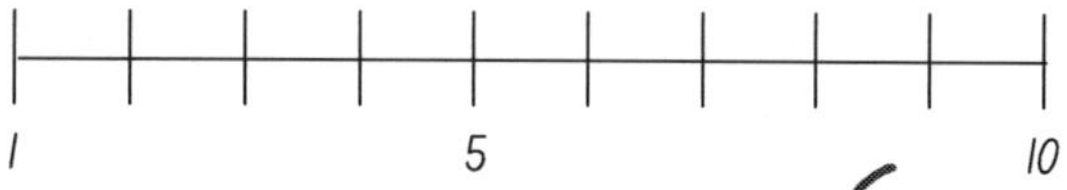

5 + 1 = 

2 + 7 = 

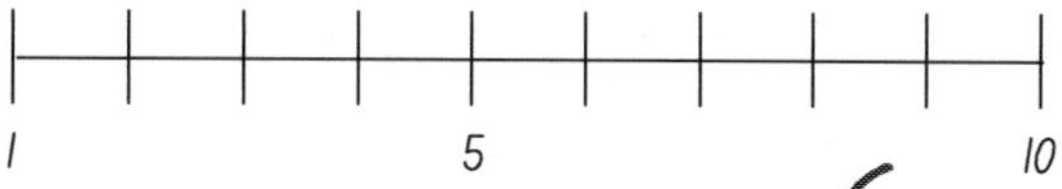

1 + 2 = 

7 + 1 = 

Name: ______________________

DIRECTIONS: TRACE THE WORDS AND NUMBERS BELOW.

four

4 4 4 4

4 4 4 4 four

COLOR FOUR HEARTS:

CIRCLE THE FOURS:

1	2	3
4	5	4
3	1	2
4	3	5
2	1	4

Name: ______________________________

SUPERHERO COUNTING

Count ther superheroes and put the correct number in the space provided.

Name: ____________________

Roll 'em and Add 'em

Add up the dice pictured. Put the answer in the space provided. Turn the page over and use your own dice to roll numbers for the equation.

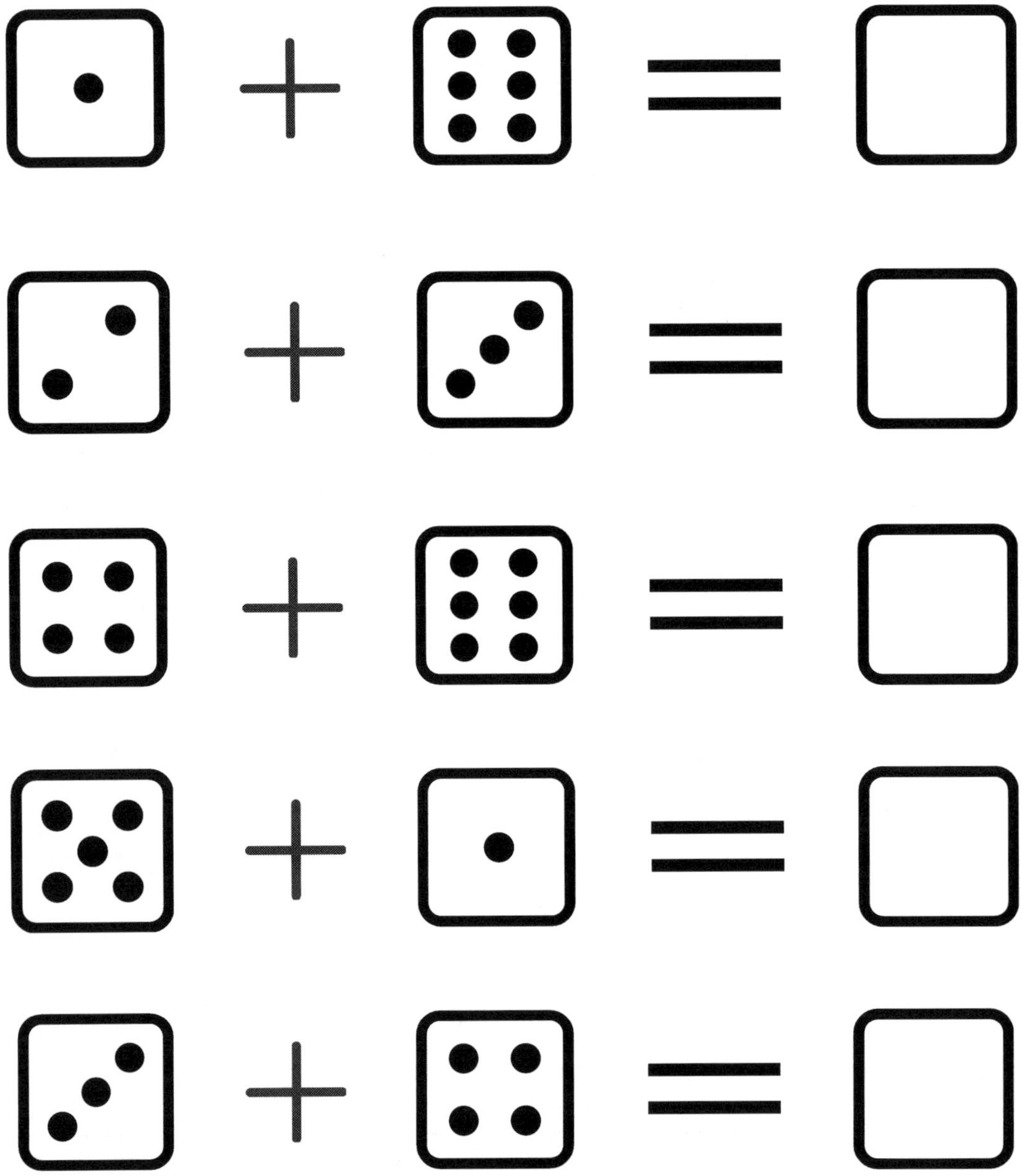

Name: ______________________________

☐ + ☐ = ☐

☐ + ☐ = ☐

☐ + ☐ = ☐

☐ + ☐ = ☐

☐ + ☐ = ☐

☐ + ☐ = ☐

Name: ______________________

☐ + ☐ = ☐

☐ + ☐ = ☐

☐ + ☐ = ☐

☐ + ☐ = ☐

☐ + ☐ = ☐

☐ + ☐ = ☐

Name	**Teacher**
Class	**Date**

Apple Subtraction

Solve the subtraction by crossing out the numbers to be subtracted. Write the answers on the line.

5 - 2 = ______

3 - 2 = ______

4 - 1 = ______

3 - 1 = ______

5 - 1 = ______

Name:

DIRECTIONS: TRACE THE WORDS AND NUMBERS BELOW.

five

5 5 5 5

5 5 5 5 five

COLOR FIVE CIRCLES

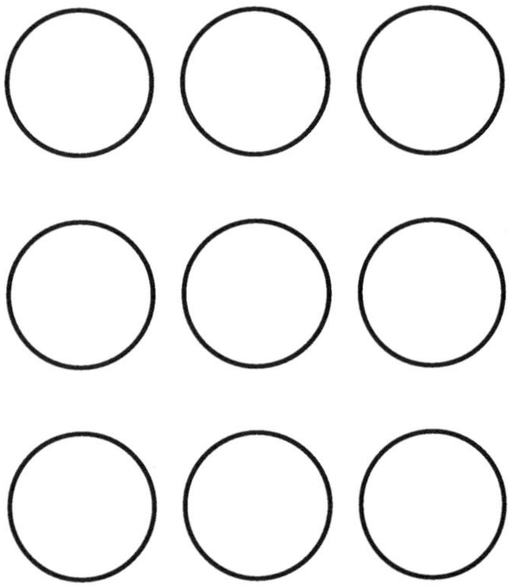

CIRCLE THE FIVES

5	2	3
1	5	4
3	5	2
4	3	5
2	5	4

Name

Class

Teacher

Date

Apple Additions

Add the apples together. Write the sum on the line.

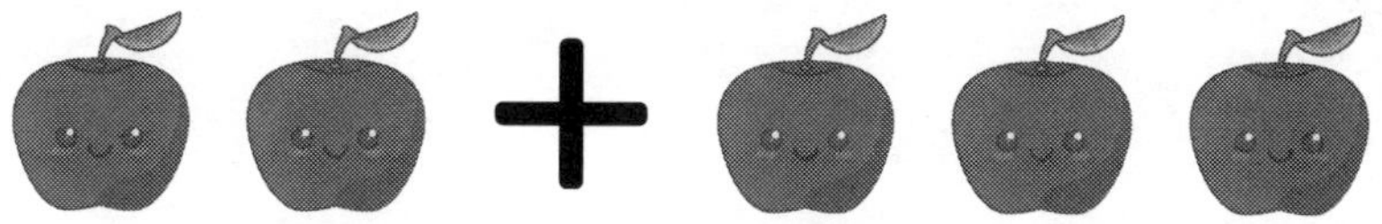

= ______

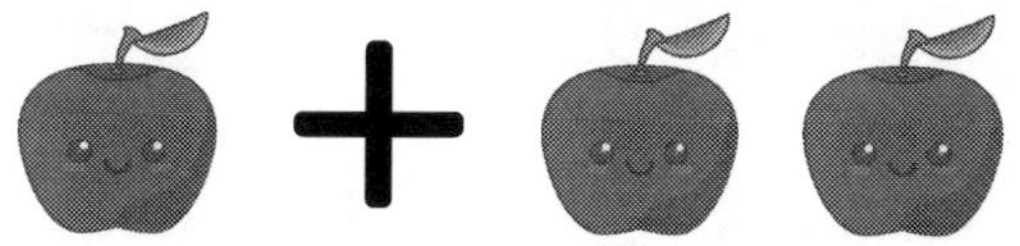

= ______

= ______

= ______

= ______

Name: ______________________________

SPRING MATH PRACTICE

Write the number that completes the equation in the blank.

6 - _ = 4	_ -1=7	5 - _ =5
3+ _ =5	5 - _ =0	5 - _ =1
5 - _ =2	_ +1=10	4 + _ =5
_ -1=6	10 - 7 = _	2+ _ =8

Name: ______________________ Date: ______________________

Directions: Help Lucky add his shamrocks!

 = ____

 = ____

 = ____

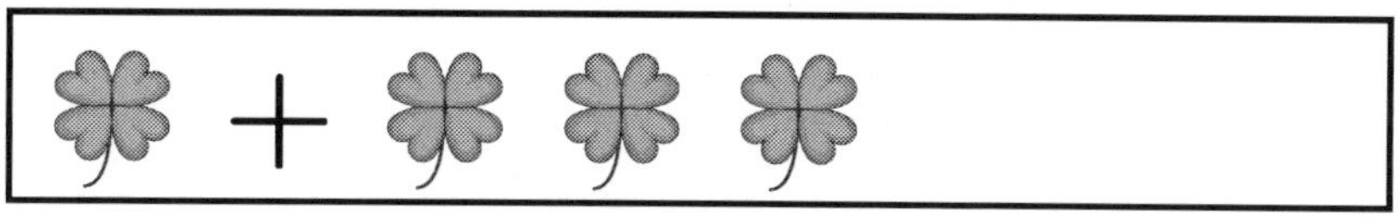 = ____

 = ____

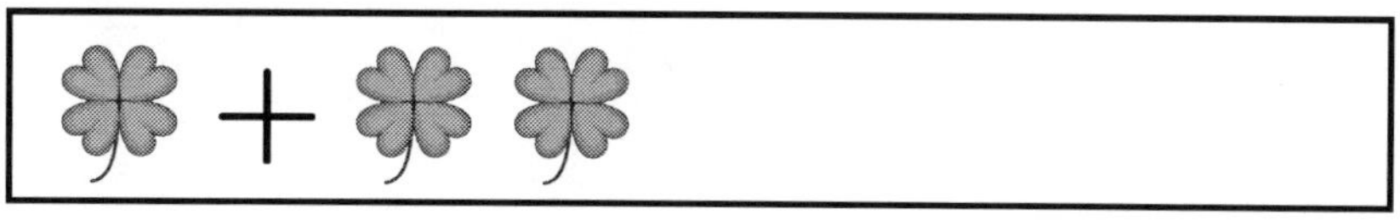 = ____

 = ____

Name: ______________________________

DIRECTIONS: TRACE THE WORDS AND NUMBERS BELOW.

two

2 2 2 2

2 2 2 2 two

COLOR TWO TRIANGLES:

CIRCLE THE TWOS:

2	3	5
4	2	1
6	1	2
1	2	1
2	1	4

Name ______________________ Teacher ______________________

Class ______________________ Date ______________________

FIND THE MISSING NUMBER

Solve the equations below.

9 + _____ = 12

2 + _____ = 8

5 + _____ = 15

8 + _____ = 16

2 + _____ = 12

3 + _____ = 9

7 + _____ = 10

9 + _____ = 11

4 + _____ = 7

5 + _____ = 13

ADDITION

Add the numbers together to find the total.

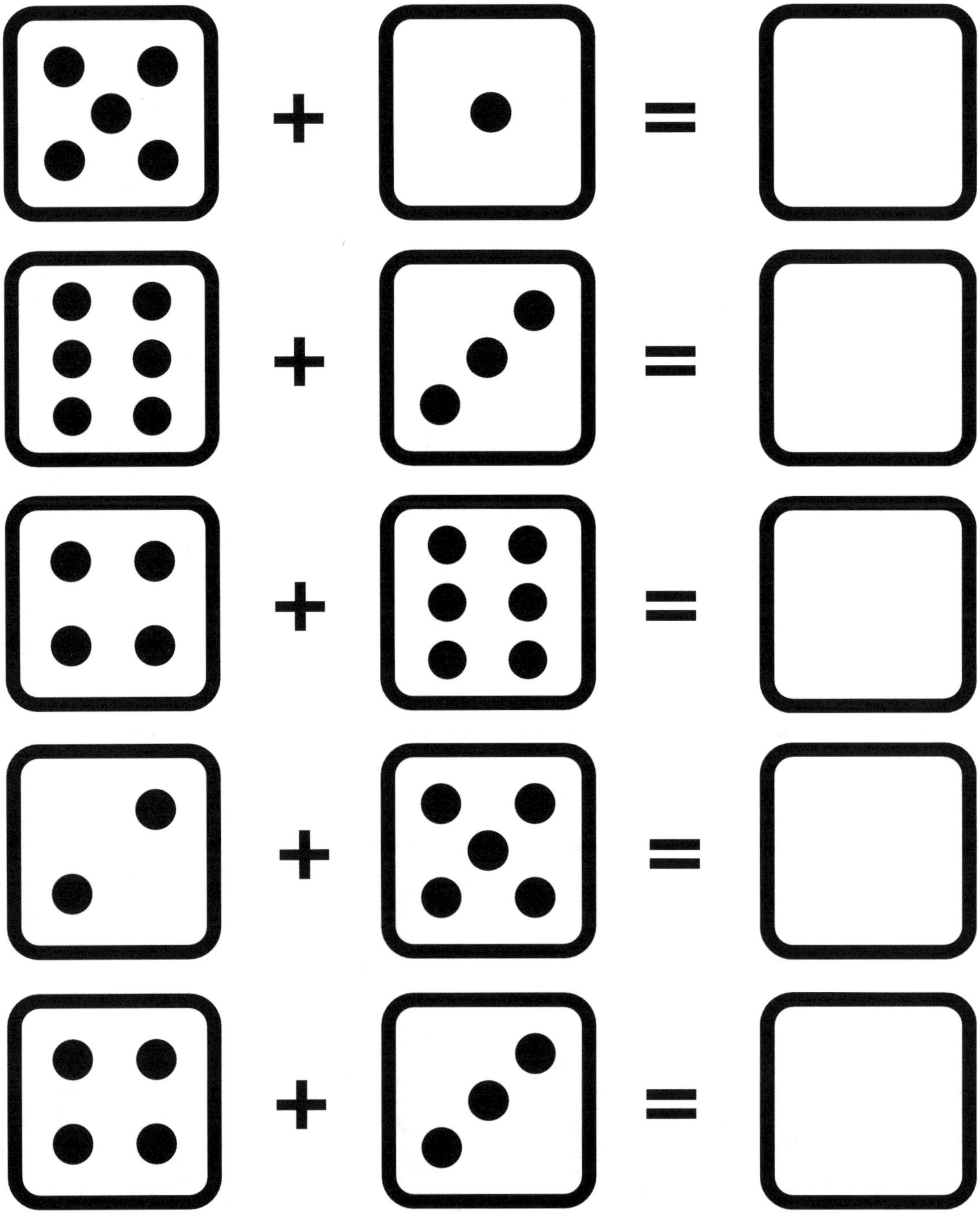

Name:		**Date:**	
Section:		**Score:**	

COUNTING CATS AND DOGS

AN ADDITION GAME

Directions: Count how many of each animal you see.
Write the answers at the box provided below.

CATS		**DOGS**	

NAME ______________________ DATE __________

SECTION ______________________ SCORE __________

Math Dice Game

Learn how to add using dice!

Directions: Add the numbers that are on top of the two dice and write the corresponding sum on the blanks.

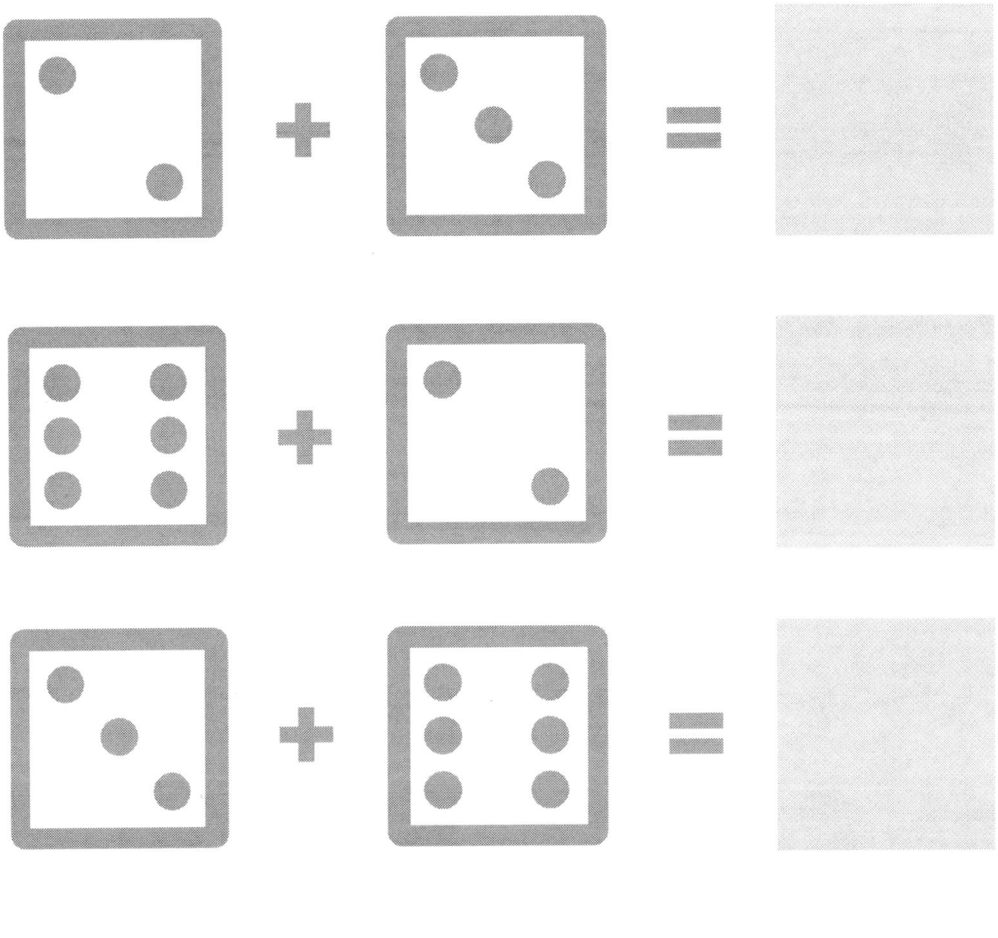

Name: ______________________________ Score _____ / 12

Stage 1 Maths: Number 1

1. Counting forwards, fill in the missing numbers: /4

10, __, 12, 13, __, 15

__, 23, 24, 25, __

2. Counting backwards, fill in the missing numbers: /4

20, __, 18, __, 16, 15

56, 55, __, __, 52,

3. Identify the number before and after: /4

__, 36, __

__, 21, __

Is it More or Less?

Learn the concept of more and less.

Instructions: Color the fruits in the right column. If there are more fruits in the right column than in the left, color them red. If there are less fruits in the right column then in the left, color them blue.

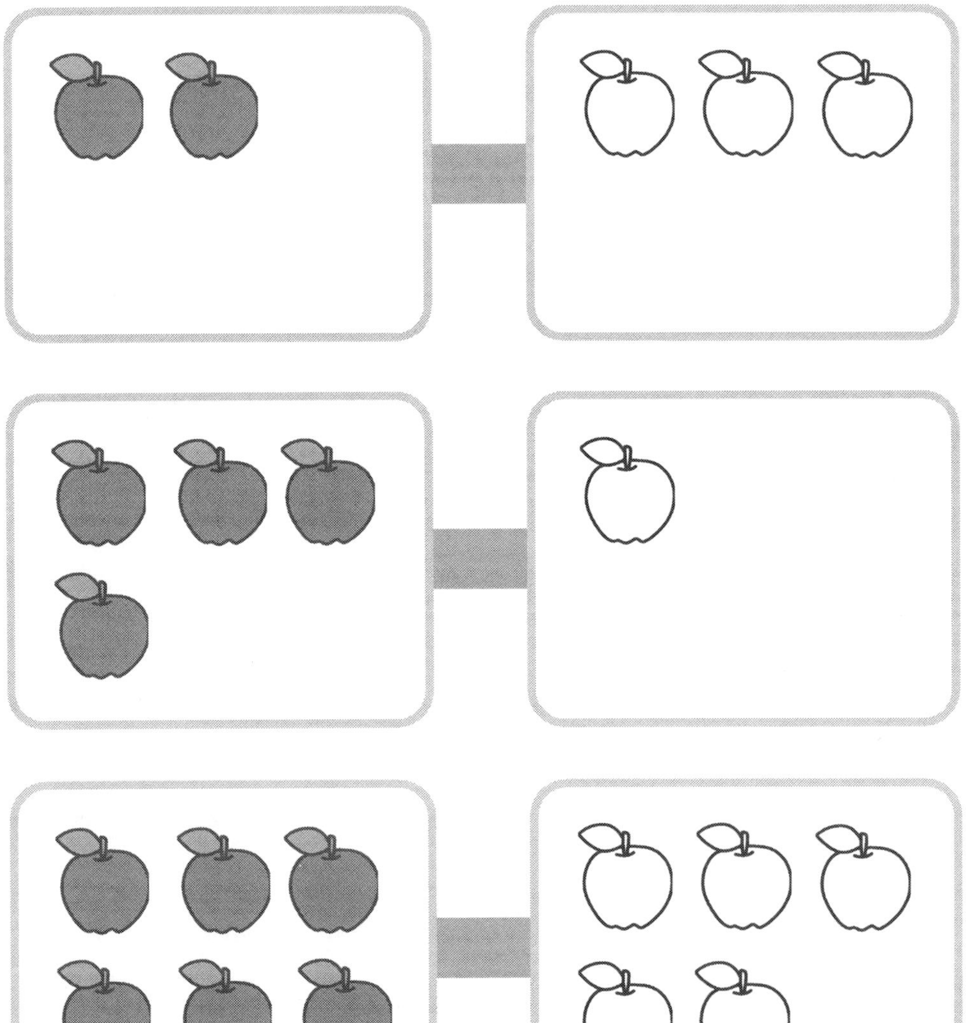

Made in the USA
Monee, IL
08 November 2026

67972900R00017